Power without Borders

Leon Neihouse

DEDICATION

To all past, present, and future members of the Dirigo Energy Institute.

CONTENTS

ACKNOWLEDGMENTS

I want to first thank the Officers (Tom Hall and Brian Spaulding) and Advisors (John Bewick, Jim Ertner, LeRoy Fournier, and Clinton Crackel) of the Dirigo Energy Institute (DEI). Without their involvement, this book would never have been written.

For chapters 3 and 4, I used the results of research projects conducted during the Spring Semester of 2009 by groups of MBA students (Scott Firmin, Susannah Levy, Joseph Rank, Andrew Sangalang for chapter 3 and Adrien Boudreau, Mariya Klevanets, Kim Clement, Patricia Shields for chapter 4) organized by Professor John J. Voyer from the University of Southern Maine.

In addition, portions of chapter 4 were derived from the CELSS concepts promoted by Terry Kok, a Perpetual Harvest Greenhouse System developed by Chris Marron, and the vertical farm work of Dickson Despommier.

James E. (Ed) Kaune worked on American Modular Power Systems, the source of the information on modular power plants found in chapter 5.

Brian O'Connell provided substantial input for chapter 6. In March of 2009 I read a Letter to the Editor concerning the cancellation of Yucca Mountain Mr. O'Connell submitted to the Brunswick, Maine Times Record. I responded and we initiated a long series of over 140 email exchanges on this topic, which continues to this day.

Don Innis and Neal A. Brown contributed the inspiration for chapter 7.

I also need to thank the Charter Members of DEI not already mentioned above - Mark C. Fitzgerald, Rick Gamache, H. Blane Howell, Stephanie B. Hathaway, Joe E. Heiss, Bob Lydon, Louis A. Marre, John Neihouse, Bernie Pinette, Vanessa L. Reynolds, R. W. Richardson, Juan Schoch, and W. A. Taylor - for their early support.

Finally, James Carberry provided invaluable assistance in unraveling the complexities of book publication, Vicente Fachina gave insights into ocean thermal energy conversion, and Carl Schlick assisted with graphics.

1. DIRIGO MEANS "I LEAD"

*I believe it is the duty of every man to act as
though the fate of the world depends on them.
Surely no one man can do it all. But, one man
CAN make a difference.* Admiral H. G. Rickover

Introduction

My professional career started with a seven year tour of duty in the
United States Navy, during which time I served in Admiral
Rickover's nuclear powered submarine program. I saw the above
quote on a 2009 calendar published by the United States Submarine
Veterans organization and I liked what he had to say.

I incorporated the Dirigo (deer-uh-go, meaning "I Lead") Energy
Institute, aka DEI, in July of 2007 and I am encouraging all DEI
members to accept the Admiral's counsel and act as if the fate of the
world depends on us. This leads to our overarching goal – research,
develop, and demonstrate methods and procedures to meet the
world's energy needs in a cost effective manner and then make this
information available for use by all national governments.

Success will mean that energy can be removed from the peace
equation - a result that could very well be instrumental in the
continued existence of civilization as we know it.

We will not be acting alone in this field. In fact, we salute and applaud all others already operating in the arena. Their actions include, but are not limited to improving energy efficiency; developing regenerative options such as geothermal, solar, biofuels, biomass, hydro, wind, waves, tides, and currents; improving fission nuclear power; demonstrating both fission (integral fast, traveling wave, and pebble bed reactors) and fusion (heavy ion, polywell, and cold) nuclear power; researching wild cards such as blacklight power and zero-point energy; and even using fossil sources such as fracked natural gas, sweet oil, and clean coal.

There is one other factor that might affect the equation – providing the energy in an environmentally benign manner.

In several billion years the sun will expand into a red giant and burn everything on this planet into a cinder. In the Academy Award winning documentary *An Inconvenient Truth*, co-recipient of the 2007 Nobel Peace Prize Al Gore implies that we are on a collision course to burn ourselves into a cinder much sooner than that. He maintains that impending global warming has possible drastic consequences for the human race, perhaps within the next 50 years. NASA professional James Hansen, PhD, with his book *Storms of My Grandchildren*, and Sir Richard Branson, with his Virgin Earth Challenge 25 million dollar prize for devising ways and means to avoid global warming, support Gore's position.

On the other side of the ledger Freeman Dyson, Professor Emeritus at the Institute of Advanced Study in Princeton, in chapter 3 *Heretical Thoughts about Science and Society* of his book *A Many Colored Glass*, takes the position that the present state of research is not solid enough to classify climate change as a civilization terminating problem. NASA professional Roy W. Spencer, PhD, with his book *The Great Global Warming BLUNDER* and Benny Peiser, PhD, Director of The Global Warming Policy Foundation, an independent Think Tank in the UK, support Dyson's position.

Without getting into a discussion on the relative merits of the two sides of this argument, we are setting a course of action under which we will be relevant under either scenario.

In order for DEI to select an energy source for development, all national governments must have access to it so as to ensure their respective countries will have power available, should they require it. This latter provision led us to adopt as the DEI motto *Power without Borders*, which is also the title of this book.

For a few brief words on the direction in which the world is heading, in 2006 Daniel Nocera of MIT wrote an article in Daedalus stating that in 2002 the world "… burned energy at a rate of 13.3 TW (terawatts) …" with a projection that "… if 9 billion people adopt the standard of living for a US resident…the world would need an astronomical 102 TW of energy in 2050." The challenge is to provide this power in a cost effective and environmentally benign manner.

MAD to MASS

Three independent nuclear warhead delivery systems - land (intercontinental ballistic missile silos), air (B-52 bombers), and sea (nuclear powered submarines with ballistic missile delivery systems) comprised a MAD policy of Mutual Assured Destruction. Any one of these three systems could independently discharge enough atomic weapons to destroy the enemy.

DEI recommends using this same methodology to formulate a MASS policy of Mutual Assured Survival Systems using a tripod offense of Solar Power, Nuclear Power, and Ocean Power. The design intent is to develop each leg of this Power Pyramid such that it will have an independent ability to supply all the world's energy needs in a perpetual, cost effective, and environmentally benign manner.

As popularized in the movie *The Da Vinci Code*, DEI is another name for God. DEI is also an acronym for Dale Earnhardt Incorporated so it should come as no great surprise that we plan to speed on our way to MASS.

Of course, speed is a relative term. On a geologic time scale, 100 years in a nine billion year life expectancy for the earth is equivalent to about 30 seconds in the life expectancy of an 80 year old human being. Our goal, therefore, is to have completed the development of cost effective and environmentally benign energy from each of these three independent sources within 100 years, or 30 equivalent seconds in the life of the earth – a very fast time indeed.

DEI RD&D (research, development, demonstration) Projects

From a practical standpoint, fossil fuel, perhaps dominated by fracked natural gas, might well be the major energy supplier for at least the rest of this century. This will undoubtedly cause increases in carbon dioxide. We are now at less than 400 ppm (parts per million) concentration and research has shown that submariners can function effectively for three month periods at up to 8,000 ppm, astronauts for three years at 5,000 ppm, and, to facilitate plant growth, controlled environment greenhouse farmers routinely raise levels to well over 1,000 ppm. Burning fossil fuels will not approach this lower level any time soon so the only potential problem in the near term is global warming.

Geothermal, biomass, wind generators, solar farms, ocean waves and tides, fission nuclear, fusion nuclear, ocean thermal energy conversion, and solar satellites in geostationary orbit are in various stages of development by many. If all national governments accept climate change as an immediate threat to the continued existence of civilization as we know it, then a global construction network coordinated by national governments can use Manhattan-type urgency and develop the above environmentally benign sources so as to solve the problem.

Unless and until climate change is accepted by all concerned to be civilization threatening, however, the course of action we will pursue is to inch along on our RD&D projects until the sources we develop are more cost effective than coal, oil, or natural gas. If we should succeed, the world's energy needs can be met in perpetuity by regenerative sources.

Our planned initial RD&D projects are:

1. A standard modular Generation IV fission nuclear power plant volume produced in shipyards and operated from floating platforms moored along navigable waterways. (Poor chance for general public support. Most people, if they accept fission nuclear power at all, will do so reluctantly. Even if all nuclear power plants are shutdown tomorrow, however, the nuclear waste problem must still be solved so this project will also investigate a method for the final disposal of spent nuclear fuel/high level radioactive wastes. The DEI proposed solution is as good as any and better than many so this aspect of fission nuclear power might meet with wide public support.)

2. An ocean thermal energy conversion plant surrounded by a floating earth based prototype of a Stanford Torus space settlement. (Good chance for general public support. Uniting these two options under one roof has never before been proposed.)

3. A solar power satellite in geostationary earth orbit. (Excellent chance for general public support. The development of this regenerative option as a practical power source would be a boon to humanity.)

4. A reduced size version of R. Buckminster Fuller's floating Triton City modified to include a four season controlled environment greenhouse. (Superb chance for general public support. Almost everyone will concur that developing a housing project self-sufficient with respect to energy/food and affordable to the average wage earner will be a good thing to have happen.)

More information on each RD&D project follows.

1. Volume producing standard modular Generation IV fission nuclear power plants in shipyards and operating them on floating platforms moored along navigable waterways:

DEI will investigate volume producing standard modular Generation IV fission nuclear power plants in shipyards followed by subsequent towing or movement by heavy lift ship to any location in the world accessible by a navigable waterway. There the plants will be moored and operated in the floating condition for an extended time without refueling or maintenance after which the complete plant will be moved to a remote facility for periodic refurbishing, refueling, and eventual decommissioning.

The latest designs that might be well suited to this operational method include, but are not limited to an integral fast reactor, the TerraPower plant promoted by Bill Gates, a Hyperion plant designed by Los Alamos National Laboratory, and MIT's design for a pebble bed reactor.

The same navigable waterway on which the plant is transported to and from its operational location can be used as a proven route to dispose of radioactive wastes from plant operation.

In the present state of the art, most reactors use less than five per cent of the installed fuel. The reprocessing of this spent nuclear fuel (SNF) can serve as one source for a future fuel supply.

Although not of significant use worldwide at this time, a SNF recycling method that DEI has targeted for inclusion in this RD&D project is pyroprocessing as it has been investigated by others, including Argonne National Laboratory. No show stoppers have yet been discovered that would prevent the implementation of this process.

Over and above this potential source, fuel can be derived in perpetuity from the four billion plus tons of uranium suspended in equilibrium in the ocean. Thus, an energy source for fission nuclear power plants will be accessible to all nations until the sun expands into a red giant.

For the final disposal of radioactive wastes, DEI will investigate a global solution to this global problem. The prime objective is to identify a final disposal method but the first order of business will be to establish a Fail Safe position by developing remote and uninhabited islands for the interim storage of all nuclear wastes. As discussed in chapter 6, this will start with low level radioactive wastes and then expand into SNF, beginning with storage in the same manner in which it is now stored at shutdown sites.

The storage capacity for SNF can be increased by using:

1. Platforms floating in protected lagoons at the storage locations
2. Some combination of double-hulled canisters and decommissioned submarine hulls to permit underwater storage in a natural or artificial lagoon at the site.

In the United States the Waste Isolation Pilot Plant (WIPP), a burial ground for military radioactive wastes, is the only disposal method now in operation. For commercial nuclear wastes, deep geologic disposal such as at WIPP is the front runner for developing a final resting place for commercial nuclear wastes. After trying for over 50 years, unsuccessfully, the United States might be able to employ findings from the Blue Ribbon Commission on America's Nuclear Future to succeed at last. However, many nations with nuclear power in their energy supply portfolio simply do not have suitable locations for deep geologic disposal.

DEI will investigate alternate methods both for nations without suitable sites and as a backup for nations investigating deep geologic disposal within their borders. Three optional targets DEI has identified are sub seabed burial, deep in the earth's magma, and ejection into space with the first option going first.

Sub seabed burial is not as "off the wall" as some might first believe it to be. It has been recommended in the past by professionals with impeccable credentials.

In her book *Trashing the Planet* the late Dixie Lee Ray, a former Chairman of the U.S. Atomic Energy Commission and Governor of the State of Washington, discussed immersing high level radioactive waste in glass, placing it in a torpedo shaped shell, and dropping it into one of the many geologically stable, remote, inactive, and non-life bearing locations on the ocean floor (so called "ocean deserts") as a suitable method to dispose of high level radioactive waste.

Bernard L. Cohen, former Professor of Physics at the University of Pittsburgh, had over 300 papers and articles as well as six books published. Dr. Cohen voiced approval of this method in a paper entitled *Ocean Dumping of High Level Waste - An Acceptable Solution We Can Guarantee* on page 162 in the Volume 47, January 1980 issue of Nuclear Technology.

But the biggest factor in favor of using this method so as to permit our nuclear wastes to Rest In Peace is Rip himself - Rip Anderson. Dr. Anderson not only gave us WIPP but also 32N164W.

As discussed in *Power to Save the World* by Gwyneth Cravens, one of many possible locations for final disposal is the ocean desert at 32N164W. At this location, a bullet shaped canister holding nuclear wastes would fall by gravity through four miles of sea water in a marine desert that ends in a 300 foot clay seabed of peanut butter like consistency with quicksand like features. The falling effect buries the canister 100 feet into this seabed with no further action required. This location is in the central area of a tectonic plate that will remain stable for millions of years.

If each canister is separated by 100 yards, the 39,000 square mile area will hold over 12 million canisters, which is enough to hold all the nuclear wastes generated by all nations for the foreseeable future.

Elements in this RD&D project might include, but not be limited to:

1. Identify remote uninhabited islands suitable for the interim storage of low level radioactive wastes, SNF, and high level radioactive wastes
2. Research the ability of breakwaters to create artificial lagoons as well as use underwater turbines, oscillating water columns, ocean waves, and wind generators so as to provide for local regenerative power
3. Develop floating platforms for placement in natural or artificial lagoons for storage of SNF and other radioactive wastes
4. Research the ability of double hulled SNF dry storage casks to provide for submerged storage
5. Investigate adapting hulls of decommissioned submarines to provide for submerged SNF storage
6. Develop an ability to transport SNF by air
7. Research floating pyroprocessing facilities to recycle SNF at its interim storage location
8. Identify and develop a location to serve as a final repository for nuclear wastes.

All national governments who have had, now have, or are planning to have nuclear power in their energy supply portfolio will be invited to participate in all aspects of this project.

For a quick summary, the plan of action is to move nuclear wastes to a selected remote uninhabited island for interim storage, possible recycling of SNF, and, after a permanent site has been developed, final disposal of all wastes - including residue from recycling, if any. The final disposal method might include, but not be limited to in a deep geologic repository, in bore holes, sub seabed burial, in the earth's magma, or ejected into space.

This fission nuclear option has one intriguing feature. Fuel can be derived in perpetuity from uranium suspended in equilibrium in the ocean and the wastes then returned to the sea from which they came.

There is a small but finite probability that a modular fusion nuclear power option, including but not limited to polywell fusion, will be developed in the near term. If it should become available, then Generation IV fission nuclear power plants need not be developed and future radioactive waste disposal will not be a consideration, except that this problem must still be solved for the nuclear wastes now existing.

2. Combining an ocean thermal energy conversion plant with an earth based prototype of a Stanford Torus space settlement:

The oceans are a vast renewable energy resource - available 24 hours a day, 365 days a year (366 in a leap year).

In many locations, the solar energy absorbed by the tropical seas create a temperature difference between the warm surface water and the cold deep water of 36°F or greater. If less than one-tenth of one percent of this stored energy is converted into electric power, it can supply all the world's electrical energy needs.

In one possible OTEC (ocean thermal energy conversion) generating process, warm sea water on the surface passes through a heat exchanger, vaporizing a low boiling point working fluid to drive a turbine, generating electricity. Cold deep sea water is then pumped up to condense the vapor, completing the power generating cycle.

The OTEC process can use the cold seawater, rich in nutrients, to support the growth of shell fish, fin fish, and other food. Additional profit sources derivable from this process include cold water for refrigeration, drinking water, and extraction of useful products suspended in sea water.

The OTEC plant can get its products to market using tankers and/or air transport. For one example of the second option, a demonstration OTEC plant could be sited in close proximity to one of many islands near the equator in the Pacific Ocean such as Howland Island whose airport, outfitted in the late thirties for use by Amelia Earhart, could be upgraded to accept modern day planes transporting workers and products to and from the OTEC plant.

A floating housing development surrounding the OTEC plant can be patterned after a Stanford Torus space settlement. The design parameters to the nearest ten feet, derived from Page 89, Space Settlements, A Design Study, NASA SP-413 of 1977, are 5,910 feet outside diameter, 430 feet torus diameter, 50 feet diameter for each of six spokes to the hub, and a 430 feet diameter hub.

The design intent is for this development to accommodate 10,000 full time residents. The people living in or visiting this city might have no intrinsic interest in DEI but the earnings before taxes of the OTEC plant, the housing development, and of all businesses located therein will be tithed for non-profit purposes, starting with DEI.

3. A network of solar satellites in geostationary earth orbit:

In the September/October '89 issue of the *Humanist* magazine, the science fiction author Isaac Asimov presented a scenario in which a network of solar power satellites in geostationary earth orbit (GEO) would provide the world with safe, environmentally benign, perpetual, and cost effective energy. Dr. Asimov suggested that this should be undertaken as a global mega engineering project to join all nations in the pursuit of a common goal.

For a potential show stopper to this project, a U. S. Department of Energy study in the 1980s, although establishing that microwaves would not be a safety hazard for people or birds, determined that photovoltaic cell production and launch costs made this method financially prohibitive. Given the dramatic reduction in the cost of solar cells and the prospects of much less expensive launch methods (electromagnetic, laser, phased microwave – for three examples) solar satellite power might now be competitive with coal, oil, and natural gas.

Placing a satellite in GEO, at just over 22,000 miles above the surface of the earth, means it will orbit the earth at the exact speed at which the earth rotates, to always remain in the same position above the equator. In this configuration, the satellite can use its photovoltaic cells to generate electricity 24/7, which can be transformed into microwaves or lasers, and transmitted to a receiver on earth. From this position, the microwaves or lasers can be transformed into electricity and either sent to customers over an electrical grid or through use of an energy carrier such as hydrogen or methanol.

Now imagine that, instead of only one satellite, there are 1,000 equally spaced satellites in GEO. Given that GEO is 26,200 miles above the center of the earth this equates to a circumference of over 164,000 miles or one satellite at each 164 mile interval. If now, over time, each of the 1,000 satellite locations is expanded into a 100 gigawatt source, then the system will be producing 100 terawatts to provide the lion's share of the 102 terawatts needed to satisfy Nocera's 2050 projection to permit nine billion people to use energy at the same rate as an average American.

Required periodic maintenance can be provided by 10 space settlements in GEO, with each accessible by a space elevator.

The full system could have a global headquarters to:

- contract for receiver and photovoltaic cell volume production
- contract for launch services
- coordinate robotic centers in space for satellite assembly
- operate and maintain the satellites
- research the use of asteroids or the moon as a source of raw materials for satellite construction
- build space elevators and space settlements.

Each of one thousand franchisees would be set up as an independent power producer to not only own, operate, and maintain assigned receivers but also sell to its customers the power generated. Global headquarters would be owned by many separate and independent entities (national governments, investment firms, and private investors) and each franchisee would be financed by private investors, utilities, and national governments serviced by the franchisee.

Energy from the solar satellites will be received on earth and then either converted to high voltage electricity for transmission over the grid or transformed into an energy carrier such as hydrogen or methanol for transportation to customers around the world. Modes of transportation might include tankers, barges, pipelines, rail cars, trucks, and service stations.

A completed network of solar satellites would realize the dream envisioned by Peter E. Glaser (invented the concept in 1968), conceptually developed by Gerard K. O'Neill (recommended their volume production to fund space settlement construction), and promoted by Isaac Asimov (called for their implementation to support world peace under an international construction program).

4. Reduced size version of R. Buckminster Fuller's floating Triton City:

In November of 1968 the Triton Foundation prepared *A Study of A PROTOTYPE FLOATING COMMUNITY* for the U.S. Department of Housing and Urban Development. The design was four acres in area, 20 stories high, elements (condos, classrooms, stores, offices, services) prefabricated on a modular basis, and created living accommodations for five thousand people.

The U.S. Navy reviewed the design and estimated that it could be volume produced in shipyards at a price that would make homes affordable to the average wage earner.

This project will revisit the design under a goal to provide a reduced size version that includes a four season controlled environment greenhouse along with insulation/renewable energy sources such that residents will be self sufficient with respect to energy and food.

This RD&D project will complement the business introduced in chapter 3 – Micro-City™ Enterprises.

Methanol as Foundation

As implied in the book *Beyond Oil and Gas: The Methanol Economy* by George A. Olah, Alain Goeppert, and G. K. Surya Prakash methanol is one way to get the power generated by OTEC plants and solar power satellites to market. Using the generated power to create hydrogen would be the first step but the next step would be to combine the hydrogen so generated with carbon dioxide to form methanol. The creation of methanol from fossil sources and a methanol distribution system is already a mature technology. A small RD&D project within the larger RD&D solar satellite and OTEC projects would be to take hydrogen from electrolysis and combine it with carbon dioxide to create methanol.

DEI Past

This is my second time around for DEI. I incorporated the first version in 1992 with a mission to conduct R&D into commercial nuclear power plants volume produced in shipyards, solar power satellites operating in geostationary earth orbit, and electromagnetic launchers lifting solar satellite parts into orbit and ejecting radioactive wastes into a sun trajectory.

During the early DEI years I sent introductory material on the DEI mission to 300 randomly selected members of the American Society of Naval Engineers. Ten joined.

In an attempt to increase the membership base I sent the below Letter to the Editor of the 50,000 member MENSA organization:

MENSA BULLETIN
Letters – October 1996 – Page 10
Global Goals: Energy

To follow up on my Letter of October '95 in which I requested proposals for global goals, I believe one worthy of consideration is solving all the world's energy problems for all time. Here is the plan:

One thousand solar-powered satellites in geosynchronous orbit (a home for former President Bush's thousand points of light) collect solar energy for subsequent transmission to receiving stations on Earth. The power is then either sent over high-voltage transmission lines, or used in the electrolysis of water to create hydrogen. This is then fed into a piping distribution network for use in generating electricity in fuel cells, or in combustion to support transportation, home heating, etc.

Uranium is continuously being washed from land to be held in the oceans in a state of suspended equilibrium. This powerful energy source can be retrieved from the sea and formed into a blanket surrounding a series of controlled micro-thermonuclear explosions. The neutron flux thereby created causes the dissociation of water into hydrogen and oxygen, and the transmutation of U-238 into plutonium. The hydrogen so generated is distributed as described above and the plutonium is used as fuel for commercial submarine nuclear power plants operating in a state of neutral buoyancy from offshore submerged concrete cocoons.

Slingatrons - first cousins of particle accelerators, magnetically levitated bullet trains and coil gun electromagnetic launchers – can be used not only to propel the high-level radioactive waste generated by nuclear power plant operation into a sun or moon trajectory, but also to launch material into low Earth orbit for solar-powered satellite construction.

The satellite and nuclear power sources operate as central power stations; these are supplemented by local power stations using such sources as geothermal plants, low-head hydro plants, solar cell farms, etc.

To prepare for an asteroid impact comparable to the one thought to have exterminated the dinosaurs, the satellite receiving stations and concrete cocoons are placed at locations with a high survivability potential. Wireless power transmission from the submarine nuclear power plants to the satellite receiving stations will ensure survival of the human race, along with its scientific and technical foundation, during the long interval of darkness.

All of the above elements, including concrete cocoons, have been discussed in reputable scientific and technical publications. The only unique concept is their union in a systematic framework. The Dirigo (deer-uh-go, meaning "I lead" and the motto of the State of Maine) Energy Institute has been incorporated to coordinate the implementation of this approach.

The processes described can continue until the sun expands into a red giant. When this happens - about 4 billion to 5 billion years from now – the Earth will be completely engulfed and all of its surface water vaporized. Life as we know it will then be extinguished and this will all become moot.

Leon J. Neihouse
24 Oak Grove Avenue
Bath, ME 04530

I received a response from only a few – none of whom volunteered to assist in the effort. One did, however, suggest that I investigate blacklight power - the release of energy through the catalytic collapse of hydrogen to below its conventional ground state. This is an intriguing concept and a candidate for future investigation.

I tried several other methods to get support for DEI, one of which was to put together a package and send it to all utilities in the United States.

No one answered so I terminated DEI in 2001.

DEI Future

Hope springs eternal. I went to a semi-retired status at my day job in April of 2007, resurrected DEI in July of that year, prepared a brochure, initiated a web site at http://dirigoenergy.org/, and wrote this book. My retirement from my day job in 2012 gives more time to devote to DEI.

The present plan for DEI has a management team ideal form of nine consisting of a President, Secretary, Treasurer and six Vice Presidents with each Vice President heading up one of the below Departments:

- Solar – an initial concentration on power satellites in geostationary earth orbit
- Nuclear – commercial land based and floating modular power plants of the fission or fusion variety
- Ocean – an initial concentration on OTEC plants
- Methanol – combines hydrogen and carbon dioxide to serve as an energy carrier
- Funding – seeks financial support

- Publicity – makes information on all DEI activities available to all.

Each of the nine members of this management team will have three assistants to arrive at 36 employees.

Some management books maintain that 40 is an ideal organizational size. The Corporate Office of DEI will arrive at this number with four Project Managers, each of whom will be assigned to one of four RD&D projects set by the DEI Board of Directors - starting with the four introduced earlier in this chapter.

The initial funding objective will be to collect pledges of $20,000 - which input will be used to submit an application to the IRS for 501(c)(3) tax deductible status and seek money to begin funding the initial four DEI RD&D projects. This pledge will be in a format patterned after the following:

The undersigned (the "Donor") agrees to donate seed money to be used for the startup of the Dirigo Energy Institute (DEI). The Donor acknowledges that DEI is a startup organization and, as such, is a speculative venture subject to all of the risks associated with the startup of any new organization, including the risk of the entire loss of the amount identified as seed money.

The Donor hereby pledges to donate the amount identified below.

This Pledge shall be binding from the date of execution until 31 December 2014 to donate all or part of twenty thousand dollars ($20,000) identified as a minimum requirement to initiate a more comprehensive fund raising campaign. DEI has the sole and absolute discretion to accept or reject this pledge.

Amount pledged as seed money for DEI = _____

Notes:

(1) The donations will be collected after twenty thousand dollars ($20,000) have been pledged.

(2) This option will remain open until 31 December 2014. If donations in this amount have not been pledged by then, DEI will cease to function as a nonprofit organization.

Executed on this the _____ day of _____ (month) in 2012/2013/2014 (circle one)

Signature: _____

Print Name: _____

Address: _____

Email: _____

To acquire a pledge to donate, send a request via email to info@dirigoenergy.org. Alternatively, join the Yahoo group at http://tech.groups.yahoo.com/group/DirigoEnergy/, go to the Files, download the pledge, and then leave the group.

After acquisition of the initial $20,000, conventional financial sources to be approached include donations from various nonprofit organizations and grants from applicable State, National, and International agencies.

To account for the possibility that DEI will not be of sufficient interest to others to donate this $20,000, a non conventional financing method will be investigated, which is the subject of the next chapter.

2. LJN ENTERPRISES

I have prepared concept information for six possible businesses. Each company will be incorporated with a perpetual and irrevocable requirement to tithe profits for nonprofit purposes, starting with DEI.

These six potential businesses, listed in the order which I think would provide the greatest benefit to the world, are:

- Micro-City™ Enterprises (MCE) - affordable housing self sufficient with respect to energy and food is at the top of the agenda.

- Micro-Farm™ Enterprises (MFE) - organic food grown in a local four-season controlled environment greenhouse is a close second.

- Allied Modular Power Systems (AMPS) - modular power is a concept of growing international interest.

- Remote Island Waste Management (RIWM) - sooner or later someone must solve the nuclear waste problem in a manner that is acceptable to the general public.

- Power Breakwater™ Enterprises (PBE) - there will always be a market niche for a source of regenerative power that creates calm coastal waters.

- Leon's Peerless Products (LPP) - a conventional business with a non-conventional feature - a source of money to be used for nonprofit purposes.

My initials of LJN will be used in LJN Enterprises, a holding company with a mission to start all of the above six companies.

Although it is expected that each business will follow a unique path for financing, each will start with a standard investment pattern that has general characteristics as follows:

- **Seed**: two hundred thousand dollars ($200,000) guaranteed in a bank line of credit (LOC) for four percent (4 %) of company stock. The LOC will be used to incorporate the company, open and outfit an office, and hire a nucleus group with the expertise necessary to refine a Concept of Operations and prepare a business plan sufficiently detailed to acquire the money necessary to continue with startup. If, after three years, the company is unable to acquire the financing to proceed with startup then these investors will be required to pay off the bank LOC and will receive for compensation ninety-five percent (95%) of the stock of the company. They will thus own ninety-nine percent (99%) of the stock to have control of the company and can realign startup operations to fulfill their expectations. If the company is successful in initial startup activities, these seed money guarantors will have an option to provide all or part of the next level of financing.

- **Estimates**: two million dollar ($2,000,000) investment for ten percent (10%) of company stock. The investment will be used to expand the nucleus group with the expertise necessary to design and prepare detailed drawings and cost estimates for a prototype and assemble a team of Joint Venture participants to support company startup. Pro forma financial statements will be developed to show profitability, which will be used to seek financing to continue with company startup. Investors will be guaranteed a three times return within five years. If, after five years, the company is unable to acquire the financing to proceed with startup then these investors will receive eighty (80%) of the stock to have control of the company and can realign startup operations to fulfill their expectations. If the company is successful in this phase of startup, these investors will be paid $6,000,000 - three times their initial investment – and will have an option to participate in the next level of financing.

- **Prototype**: two hundred million dollars ($200,000,000) for fifteen percent (15%) of company stock. The investment will be used to expand the staff, build and operate a prototype, and replicate the prototype as dictated by operational success. Investors will be guaranteed a five times return within seven years. If, after seven years, the company is unable to acquire the financing to proceed with startup then these investors will receive sixty percent (60%) of the stock to have control of the company and can realign company operations to fulfill their expectations. If the company is successful in this phase of startup, these investors will be paid one billion dollars ($1,000,000,000) - five times their initial investment – and have an option to participate in an Initial Public Offering (IPO).

As proposed above, each company will have a maximum of 15 years to get through the IPO stage. If successful then its stock will be held as follows: one percent (1%) by the initial participants, five percent (5%) by the Estimating Team, five percent (5%) by the Prototype Team, twenty nine percent (29%) by the investors, and sixty percent (60%) sold in the IPO and/or held by company employees.

Thus, unless and until at least one of these business concepts reaches a mature form, DEI is likely to remain a small and unnoticed skeleton organization – for a time span extending up to 15 years.

A standard pledge to provide the seed money for each company proposed in this chapter will have characteristics as follows:

Seed Money Pledge for (Name of Company)

The undersigned (the "Investor") agrees to guarantee all or part of a bank line of credit (LOC) to be used as seed money for (Name of Company), a business proposed by LJN Enterprises.

The Investor has read the Concept of Operations for (Name of Company) and has had discussions with other persons and sources as deemed reasonable and appropriate before deciding to guarantee this seed money.

The Investor acknowledges that (Name of Company) is a startup company and, as such, is a speculative venture subject to all of the risks associated with any startup business, including the risk of the entire loss of the guaranteed amount.

The Investor hereby represents, warrants, and covenants to have the financial wherewithal and capacity to provide for his or her personal needs and contingencies without regard to the amount guaranteed, can afford a complete and entire loss of this amount,

and will have no need in the foreseeable future to use this guaranteed amount for any other purpose.

The Investor hereby agrees to guarantee the amount identified below. With the execution of this pledge and after incorporation of (Name of Company) eight thousand shares (8,000) will be transferred to the investors and one hundred ninety thousand (190,000) of the two hundred thousand (200,000) shares authorized for (Name of Company) will be held in an escrow account. In the event the investors are required to pay off the LOC, this stock will be transferred to the investors in proportion to the amount guaranteed such that the investors as a group (200 guaranteeing $1,000 each; one guaranteeing the entire amount; or any combination to equal $200,000) will own ninety nine percent (99.0%) of (Name of Company). In the event the investor does not have to pay off the LOC, he or she will be given an option to participate proportionately in the next level of funding necessary to continue with the startup of (Name of Company).

This pledge shall be binding from the date of execution until 31 December 2014 to guarantee a two hundred thousand dollar ($200,000) three-year bank line of credit. LJN Enterprises has the sole and absolute discretion to accept or reject this pledge.

Amount guaranteed as seed money for (Name of Company), in increments of $1,000 equals _____ .
Stipulations, if any, are attached.

Executed on this the _____ day of _____ (month) in 2012/2013/2014 (circle one)

Signature: _____

Print Name: _____

Address: _____

Email: _____

Each business will tithe profits to one or more nonprofit organizations selected by LJN Enterprises. As earlier noted, DEI will be the first recipient of these donations.

I will send a Non-Disclosure Agreement and a pledge to invest in any of the six businesses discussed in this book to those financially qualified readers who find one or more concepts of interest. After this, I will send a proprietary Concept of Operations and will then respond to questions by email, by phone, or in person.

If and when $200,000 are pledged for a business, then a telephone or face to face meeting will be set up to settle the details required to guarantee the money and proceed with business startup.

Each of the following six chapters gives more information on one of the six business concepts.

3. MICRO-CITY™ ENTERPRISES

"You never change things by fighting the existing reality. To change something, build a new model that makes the existing model obsolete." R. Buckminster Fuller

This proposed business is of special personal interest to me. My last name of Neihouse is an Americanization of the German name Neuhaus, meaning new house. I was born and raised on a farm so I seem destined to get involved in new housing developments that have a self contained farm – or, in this case, a four season controlled environment greenhouse.

The objective of Micro-City™ Enterprises, or MCE for short, is to design small housing developments that are made affordable by volume producing standard designs for a floating or land based configuration and then placing them in urban, suburban, and country settings.

Long ago R. Buckminster Fuller offered one solution to the affordable housing problem – at least in principle. In November of 1968 he published *A Study of A PROTOTYPE FLOATING COMMUNITY*, known as a Triton City. Fuller demonstrated that, if volume produced in shipyards, this floating design will be affordable to those being paid a minimum wage.

The intent of MCE is to make developments such as this energy self sufficient through the use of techniques to include, but not be limited to insulation, below grade heat sources and sinks, waste to energy, solar, wind generators, ocean waves, and ocean currents.

Waste to energy, solar, and wind are available for use in both land based and floating configurations.

In addition, energy self sufficiency in a land based housing development can use regenerative energy such as geothermal by the simple expedient of building in close proximity to such a source.

As discussed in chapter 7 of this book, a floating housing development can rest in calm waters created by a power breakwater and have its energy supplied by this very same power breakwater employing ocean waves, winds, and currents as regenerative sources.

For food, an integral part of the housing development will consist of a Micro-Farm™, such as one of those discussed in the next chapter of this book. It will be a four season controlled environment greenhouse that will use NASA developed CELSS (Closed Ecological Life Support System) concepts. These will adjust carbon dioxide levels, light type and intensity, organic nutrients delivered via a hydroponic and/or aeroponic process, temperature, and moisture content under a design objective to achieve maximum growth with minimum waste discharge to the environment for a product line selected by the homeowners. Products not claimed by the homeowners will be sold to walk in customers or retail outlets.

One housing development design, in the shape of a Mayan Pyramid, will have a greenhouse farm on the sun facing side complemented with condos on the other three sides. All homes will open outward to an exterior privacy patio and inward to a community center. The greenhouse will use terraced grow beds in a configuration starting at ground level and rising to a rooftop setting to provide products selected on a periodic basis by the homeowners.

Other design concepts are floating or land-based reduced size versions of three Buckminster Fuller designs (Tetrahedron City with 1,000,000 people, Old Man River City with 125,000, and Triton City with 5,000), an Arcosanti design, geodesic and Fly's Eye domes of several sizes, and various designs of energy efficient single and multiple family homes.

So as to obvert the necessity of owning an automobile, shared transportation will be available in each development. This will include both a leasing program for automobile short term use and a microbus making periodic trips from the development to grocery stores, shopping malls, an airport, a bus terminal, a train station, and entertainment venues.

The long term objective of this proposal is to build these Micro-City™ complexes using at least 10 standard design concepts. The overarching purpose is to volume produce each standard design such that the development is energy and food self-sufficient with wastes recycled to the extent practical. It should be mentioned that the designs will be standard in much the same way that the design of the human body is standard. There will inevitably be as much variety in different housing developments of the same design as there is variety in different human beings of the female or male design.

A possible international marketing campaign could employ a cruise ship embarking on a world tour entitled *The Voyage of the Micro-City™ AEGIS*, where AEGIS is an acronym for Amalgamated Entertainment, Games & International Shopping.

An existing cruise ship can be refurbished in the form of a floating housing development to contain a greenhouse, penthouse and luxury staterooms, standard staterooms, crew quarters, a shopping section with retail outlets from world class companies, an entertainment section with state-of-the-art facilities, and a section featuring traditional carnival games and rides.

The tour, with stops in ports around the world, will attract people with a combination rock concert and carnival atmosphere (The circus is coming to town!).

The lion's share of the profits from the tour will be used to begin the construction of selected housing developments in the host country to provide health, housing, energy, food, clean air, and fresh water to shield (derived from the AEGIS name) the poor and homeless in the outlying areas from the effects of poverty through a life style traditionally inaccessible to them.

To summarize, the overarching purpose of this strategy is to seek energy and food self-sufficiency in a neighborhood venue. Key elements are:

- Volume produce standard structures so as to create a lifestyle affordable to all
- Provide energy from regenerative sources
- Grow food in a CELSS environment that permits a four-season output
- Employ Fuller's concept of a Design Science Revolution to provide continuous product improvement in energy and food.
- Use *The Voyage of The Micro-City™ AEGIS* to market MCE on a global basis.

Each Micro-City™ will be organized as a franchisee of MCE. In an ideal form, standard operating procedures in these micro-cities organized on an international basis will permit easy homeowner transfer between cities so as to permit individual homeowners to find a home and community design suitable to their life style. As with the other business concepts, 10 % of the earnings before taxes not only from each franchisee but also from corporate MCE headquarters will be donated to non-profit causes, starting with DEI.

A Joint Venture team to develop this concept might consist of expertise in:

- Housing development design
- Greenhouse design
- Floating configuration design
- Floating energy sources
- Land based configuration design
- Land based energy sources
- Real estate
- Legal
- Marketing.

This Joint Venture team will produce the marketing materials that are expected to result in customers on a global basis.

Seed money will be required to incorporate MCE, open and outfit an office, and hire a nucleus group with the expertise necessary to define many model options and prepare a business plan sufficiently detailed to acquire the money necessary to expand the nucleus group so as to assemble a Joint Venture team, define many model options, select one model for a world tour, and prepare detailed drawings for the model and cost estimates for both the model and the tour.

More money will be needed to expand the staff, build the model, conduct the world tour, and get contracts for at least one development in each country in close proximity to each stop.

For a Non-Disclosure Agreement and a pledge for Micro-City™ Enterprises send a request via email to info@dirigoenergy.org. Alternatively, join the Yahoo group at http://tech.groups.yahoo.com/group/DirigoEnergy/, go to the Files, download the pledge, and then leave the group.

The next chapter will give more information on a greenhouse that will be an integral part of each Micro-City™.

4. MICRO-FARM™ ENTERPRISES

A growing global population must eat so agriculture is trending towards large vertical farms in urban areas to feed the local population - a concept more fully explored in the book *THE VERTICAL FARM Feeding the World in the 21st Century* by Dickson Despommier. It is easily possible to extrapolate this trend downwards so as to place small four season controlled environment greenhouses in neighborhoods at the Micro-City™ size.

Micro-Farm™ Enterprises (MFE) will investigate the profitability of this concept through a franchise company under which franchisees will use four season controlled environment greenhouses that recycle wastes using CELSS (Closed Ecological Life Support System) techniques to grow a product line that might consist of, but not be limited to vegetables, herbs, flowers, fish, fowl, rabbits, and pigs; a reduced product line; or single crops such as lettuce, grapes, strawberries, tomatoes, cantaloupes, watermelons, or medicinal plants. Each greenhouse will sell its products to walk in customers and/or to retail outlets.

The greenhouse will optimize light, temperature, carbon dioxide (CO_2) enrichment, air moisture content, and soluble nutrient levels in conjunction with continuous planting and harvesting. Year-round operation will permit continuous growth of vegetables, herbs, fruit, flowers, fish, fowl, rabbits, and pigs. One crop of potential interest is grapes grown under controlled conditions that replicate the environment that existed during the years when award winning wines were produced.

Growing techniques will provide artificial light, CO_2, and soluble nutrients absorbed through roots and leaves. The system enhances growth by proportionally increasing light, heat, water, CO_2, and nutrients at specified times of the day.

A standard greenhouse in the northern hemisphere might have a tall northern wall with planting beds stacked upward toward the northern wall. From an end view, the greenhouse will appear similar to an A-Frame. In a standard layout the floor of the greenhouse will be terraced for vertical growing. The length of the greenhouse will be determined by the franchisee, sizing it to desired production levels.

Besides terraced beds, it will be possible to apply the verti-grow method that utilizes pots stacked one above the other. It will also be possible to build the terraces out of enclosed fish tanks, thus allowing fish to be raised (aquaponics) as another income stream.

The open space in the A-Frame interior might house elements including, but not limited to plumbing, pumps, aquaponics, mushroom beds, composting, water and nutrient storage, an insulated nursery, and a retail outlet. A company owned model will have classrooms in this section for instructing franchisees in the design and operation of the greenhouse.

Growing techniques might include:

- **Light:** Experience indicates that approximately 11-12 hours is optimal daylight length for most common food plants in temperate zones. Day length is adjusted with artificial sunlight (LED grow lights) such that plants receive the same amount of light throughout the year.

- **Carbon Dioxide Enrichment:** Normal atmospheric CO_2 concentration is at less than 400 ppm but experience indicates that some plants prefer up to 2000 ppm. A flame (propane or natural gas) can serve as a peaking CO_2 source with baseline CO_2 levels provided by decomposing compost or other natural sources.

- **Soluble Nutrients:** Organic nutrients can be passed through a soil-less growing medium comprised of perlite, pumice, vermiculite, and decomposing organic matter (potting soil). Soluble nutrients of organic compost can be applied to the underside of leaves as a fine mist during the light/CO_2 enrichment period.

- **Others:** Additional growing techniques under evaluation include, but are not limited to water treatment systems.

The greenhouse will be designed to regulate temperature using thermostats, timers, and/or programmable controllers, all with the option for manual override. The energy management systems are operated with the intent of maintaining the desired greenhouse temperature and humidity with the minimum energy input.

NASA recycling concepts from CELSS to provide food for settlers of Mars will be incorporated into the design of these greenhouses. With this feature supplementing vertical farms and conventional farming techniques, a sustainable life support system can be built to accommodate perhaps up to 10 billion people on this planet.

The MFE methods and features include:

- Continuous year-round growing and harvesting of high quality vegetables, herbs, fruit, flowers, fish, fowl, rabbits, and pigs
- Ability to grow 'designer' fruits and vegetables

- Reduced need for pest control due to compost based nutrient application keeping soil borne insects and diseases out of the greenhouse biome
- Higher plant brix (sugar) levels, resulting in better taste and longer produce shelf life
- Maximized sunlight harvesting through use of tiered beds
- Integrating renewable energy systems to establish greenhouse energy self-sufficiency
- Significantly reduced shipping costs by raising food crops locally.

To start operations, MFE will develop company models for operation in a MFE franchise business network for placement on land unsuitable for conventional farming.

A marketing plan will employ *The Voyage of the Micro-City™ AEGIS* (Amalgamated Entertainment, Games & International Shopping) - a ship discussed in the previous chapter - that will be outfitted for travel on the high seas with built in greenhouses, restaurants, condos, entertainment, games, and shopping venues. It will embark on a world tour with a commitment to use part of the profits to build a greenhouse in each host country with provisions to grow food that will help to shield the poor and homeless in that country from the effects of poverty.

MFE will teach integrated farming and waste management techniques necessary for organic food growth in a micro-farm and also perform R&D into features designed to seek energy self sufficiency such as insulation, below grade heat sources and sinks, waste to energy, solar, and wind generators.

Experiments in carbon dioxide levels, light type and intensity, organic nutrients delivered via a hydroponic and/or aeroponic process, temperature, and greenhouse moisture content will be conducted under a design objective to achieve maximum plant growth.

Aquaponics and livestock interactions will be addressed under which air, water, and wastes are recycled so as to maintain a fresh air and clean water atmosphere.

Greenhouse dimensions will be selected such that an optimum mix of vegetables, fruits, and meats can provide all the food required for a neighborhood sized in accordance with Micro-City™ parameters.

The design intent is to volume produce an economy model for the developing world and a standard model for the middle class with custom models built for the rich and famous such that basic life support needs will be available in principle for all people on Earth.

The MFE top ten reasons to develop four season controlled environment greenhouses – listed in reverse order with apologies to David Letterman - are:

10. Prerequisite for extended trips to the Moon, Mars, and beyond

9. Reduces fossil fuel use through local food production

8. Conserves water

7. Recycles wastes to minimize release to the environment

6. No crop failures due to droughts, wind, rain, snow, hail, or global warming

5. Marginal farmland and urban rooftops used for food production

4. Nutrient rich food grown without artificial herbicides or pesticides

3. Land efficiency - one greenhouse acre equivalent to up to 30 outdoor acres

2. A continuous harvest 365 days a year (spring, summer, fall, and winter)

1. Promotes sustainable living

Seed money will be required to incorporate MFE, open and outfit an office, and hire a nucleus group with the expertise necessary to define a standard greenhouse and prepare a business plan sufficiently detailed to acquire the money necessary to expand the nucleus group so as to prepare detailed drawings and cost estimates to adapt the standard model for the world tour.

More money will be needed to expand the staff, build the prototype, conduct the world tour, and get contracts for at least one MFE franchisee in each country in close proximity to each stop.

For a Non-Disclosure Agreement and a pledge for Micro-Farm™ Enterprises send a request via email to info@dirigoenergy.org. Alternatively, join the Yahoo group at http://tech.groups.yahoo.com/group/DirigoEnergy/, go to the Files, download the pledge, and then leave the group.

The next chapter will give more information on standard modular power plants volume produced in shipyards.

5. ALLIED MODULAR POWER SYSTEMS

For quite some time I have been promoting the volume production of modular power plants in shipyards.

Between April 2, 1990 and August 22, 1991 I sent three unsuccessful submissions to the Department of Energy on this general concept as it relates to nuclear power plants.

The late Lawrence M. Lidsky, Professor of Nuclear Engineering at MIT, was instrumental in organizing an International Workshop on June 17-19, 1991 at MIT on a land based modular high temperature gas cooled reactor, or MHTGR as it was then called. I was in attendance and used the results of this workshop, as well as a concept sketch and description of a floating plant created for me by Dr. Lidsky, in a business plan that discussed volume producing the MHTGR in manufacturing facilities such as shipyards, towing the completed plants to sites excavated to accept them, and operating them in the floating condition. The business plan discussed a boundary condition under which it would be possible to volume produce the plants within the existing global shipbuilding infrastructure such that 10,000 plants could be built within 40 years.

There were too many unresolved licensing issues so all attempts to continue with business development at that time were halted but it was soon thereafter that I started the Dirigo Energy Institute. DEI received a bit of publicity through an article entitled "*10,000 nukes in the sea*" which appeared on page 9 of Maine Times, November 20, 1992. An excerpt from this article reads as follows:

The Dirigo Energy Institute, a nascent research and development group in Bath, believes the world's future energy needs can be met by floating nuclear reactors.

Floating, gas-cooled nuclear power plants, argues Institute President Leon J. Neihouse, would be "safer and cheaper" than land-based, water-cooled nuclear power plants.

Neihouse, who works for John J. McMullen and Associates, says sea-going reactors will be safer than land-based plants because they will be earthquake-proof and can be towed to remote sites for repair and refueling.

Asked whether he anticipates public resistance in light of the ongoing debates over the disposal of nuclear waste, Neihouse says that's not the Institute's problem.

"We are a research and design institute," says Neihouse. "The waste disposal problem isn't a technical problem, it's a political problem. Geologic burial methods have been proven and the plant itself can store 15 to 20 years worth of waste."

Neihouse believes building nautical nukes is a natural defense conversion industry for U.S. shipyards in the post-Cold War era.

"An ideal form of ten thousand generating sites, each holding one or more 200-megawatt plants, can be built in 40 years using the existing shipbuilding infrastructure," states a Dirigo Energy Institute press release. (EAB)

The same shipbuilding conditions exist today as in 1992 so there should be no insurmountable problems with volume producing modular nuclear power plants in shipyards except that the nuclear waste disposal problem, a subject discussed in the next chapter of this book, must first be solved.

That early version of DEI received more publicity through a Letter to the Editor that appeared in the Maine Coastal Journal of April 26, 1995. Excerpts from that **Readers Speak Out** letter are:

In a recent survey, members of the League of Women Voters ranked nuclear power as the most dangerous activity or technology. It matters little that "experts" responding to the same survey put nuclear power in the number 20 position. The inescapable conclusion is that attempts to build and operate nuclear power plants in the United States will be met with strident opposition.

In the emotional world in which we live, the logical course of action that will clean up the environment faster and save lives more quickly is for nuclear forces to concede to the anti-nuclear groups their Pyrrhic victory and then begin looking for alternate solutions to our energy problems. To start a dialogue along these lines, my personal top ten choices to develop energy sources, from bottom to top with apologies to David Letterman, are:

10. Nuclear. The tenacious opposition of the anti-nuclear movement is its Achilles' heel.

9. Oil. The little that's left should be reserved for non-energy related purposes.

8. Solar. Father Sun, our constant fair weather companion, deserts us during dark times.

7. Wind. When the doldrums set-in, set-up on Capitol Hill.

6. Hydro. It's a "Damn Sight" better than most others, but there aren't enough dam sites.

5. Coal. Enough to last for hundreds of years - after it has cleaned up its act.

4. Natural Gas. Very attractive - but Mother Earth might get hot under the collar.

3. Motion of the ocean. Tides, waves, and currents give "Mo Power to You."

2. Biomass. Garbage In, Joules Out.

1. Geothermal. Mother Earth will always be there for you.

In other words, there is no "one" solution to our energy problems. We will need small geographically dispersed plants and large central sites generating safe, environmentally benign, perpetual, and cost effective energy using a variety of mechanisms based on land, ocean, solar, and nuclear sources.

Leon J. Neihouse; The Dirigo Energy Institute, Inc.; Bath

I soon abandoned hope for the Dirigo Energy Institute in the conditions as they then existed and became a Principal in American Modular Power Systems (AMPS). This business was incorporated to volume produce gas turbine power plants in shipyards and operate them in a franchise independent power producing network that would sign BOOM (build, own, operate, and maintain) contracts to provide electricity to customers. We tried for several years but could not find financing to proceed so we folded the company.

This new attempt to volume produce standard modular power plants in shipyards will have the same acronym but it will stand for Allied Modular Power Systems and be set up as a franchise network of independent power and water producers with, as before, the network operating under BOOM contracts to build, own, operate, and maintain power plants that sell electricity as a primary product and water as a secondary one.

Each franchisee will have a mission to fulfill a Power Purchase Agreement (PPA) held by corporate AMPS headquarters. For those customers requiring fresh water for drinking and/or agricultural purposes, AMPS corporate headquarters will satisfy the requirements of a Water Purchase Agreement (WPA) by designing, building and owning standard desalination plants to be operated and maintained by the franchisees.

AMPS will start by using natural gas as the energy source for gas turbine power plants. To the extent possible, they will be volume produced in shipyards and operated from floating platforms moored at generating sites on a coastline or along a navigable waterway.

Suppliers might include, but not be limited to:

- **Gas turbines:** Pratt & Whitney, General Electric, Siemens and Rolls Royce
- **Shipyards:** South Korea, China and Japan each have the infrastructure in place to volume produce standard floating gas turbine and desalination plants for subsequent transport by heavy lift ship to any site on the planet accessible by a navigable waterway. All developed countries have shipyards to volume produce standard plants for operation in their country.

If climate change should prove to be an intractable problem, the gas turbines can later be adapted to burn methanol derived from various environmentally benign sources (ocean thermal energy conversion plants, solar satellites in geostationary orbit, large fission or fusion nuclear power plants, etc). Alternatively, the entire gas turbine power plant can be replaced with a standard modular floating Generation IV fission or fusion nuclear power plant. After the nuclear waste problem is solved and when fission/fusion nuclear becomes more cost effective than gas turbines, this will be the natural flow of events.

The standard AMPS franchisee will operate and maintain a nominal 100 megawatt power plant as well as a desalination plant, if so requested by the customer. In its ideal form, AMPS will replicate this design at 10,000 locations and thus be in a position to generate one (1.0) terawatt of power.

To put this in perspective, the present total energy requirements for the world are in the 17 terawatt range. The balance of the power required will be provided by others, using conventional fossil (coal, oil or natural gas), fission or fusion nuclear, and many regenerative options (hydro, geothermal, biofuels, wind, solar, ocean thermal energy conversion, etc).

In the ideal form, a corporate AMPS headquarters will design, market and own the power and desalination plants; 10 construction offices will monitor and oversee the companies building the standard plants; and each of 100 offices will provide life cycle support to 100 assigned franchisees that operate and maintain the plants.

Each franchisee will pay corporate AMPS headquarters:

1. An initial franchise fee
2. A set amount for waste disposal
3. A set amount for plant decommissioning
4. A set amount for corporate support
5. Ten per cent of earnings before taxes.

Seed money is necessary to assemble a Joint Venture team to consist of AMPS and expertise in:

1. Breakwater design to create calm waters for the power plant
2. Gas turbine modular floating power plant design

3. Floating platform for the modular power plant
4. Modular desalination plant design
5. Floating platform for the modular desalination plant
6. Balance of plant for generating site
7. Marketing.

The intent of this team is to support an international franchise network of independent power producers specializing in the design, construction and operation of floating modular gas turbine power plants offering a modular desalination plant option.

This Joint Venture team will produce the marketing materials that are expected to result in AMPS signing PPAs and WPAs with customers on a global basis.

The intent is to sign PPAs/WPAs without identifying a specific power source. Thus, each franchisee can decide the point at which it might transfer from a gas turbine to a modular nuclear power plant, for example. It is expected that the time at which the transfer occurs will be driven by the expectations for long-term success of the stockholders of the franchisee.

Seed money will be needed to incorporate AMPS, open and outfit an office and hire a nucleus group to assemble a Joint Venture team and prepare a business plan sufficiently detailed to acquire the money necessary to continue with startup.

Subsequent steps will be to expand the staff with the expertise necessary to acquire concept sketches for floating power and desalination plants, cost estimates for the same, concept sketches and cost estimates for balance of plant at the operating site including a floating breakwater, cost estimates for franchisee startup expenses, pro forma financial statements to show profitability for franchisee and franchisor, acquire the first PPA/WPA to build, own, operate and maintain a power plant with a desalination capability and acquire financing to continue with AMPS startup.

Additional steps will be to further expand the staff, design and build a prototype power/desalination floating plant, outfit a generating site to accept the floating plant and acquire PPAs/WPAs for nine more power plants with a desalination option. Finally, an initial public offering will provide the funds to continue in pursuit of the ideal form.

For a Non-Disclosure Agreement and a pledge for Allied Modular Power Systems send a request via email to info@dirigoenergy.org. Alternatively, join the Yahoo group at http://tech.groups.yahoo.com/group/DirigoEnergy/, go to the Files, download the pledge, and then leave the group.

The next chapter will give more information on radioactive waste disposal.

6. REMOTE ISLAND WASTE MANAGEMENT

As validated by the earthquake, tsunami, and meltdown in Japan, an untold number of unexpected events can occur with nuclear power plants. Among these are natural disasters (earthquakes and hurricanes), operator error (Chernobyl and Three Mile Island), and acts of sabotage.

One example of an event falling in the latter category would be to fly an airplane in Kamikaze fashion into a spent nuclear fuel (SNF) target of opportunity. This type of incident has yet to happen at a nuclear power plant but it has a well publicized history elsewhere. Witness the terrorists on 9/11 flying into the twin towers, a high school student into an office building in Florida, and a distraught tax payer into an IRS building in Texas. Filling the airplane with explosives would magnify the damage.

In that it is impossible to guarantee that there will be no unexpected events associated with SNF storage, a logical course of action is to place it in a location such that any and all incidents will not endanger the health and safety of the general public.

Real estate agents are fond of repeating the mantra "*Location, Location, Location*" when discussing the three most important characteristics to look for in purchasing a new home or business. An appropriate mantra might be *"Fail Safe, Fail Safe, Fail Safe"* as the three most important characteristics in finding locations to first store SNF and then provide for its final disposal.

When compared to other dangers we face routinely, the effects of low level radioactive wastes are minimal but the general public does not understand radiation, they fear it inordinately, and they do not want radioactive wastes in their back yard. This set of circumstances leads to a persistent NIMBY (Not In My Back Yard) response in all locations in which attempts are made to store or provide for final disposal of any type of radioactive wastes. In that every place in the continental United States is in someone's back yard, remote and uninhabited islands might be better temporary storage locations for SNF.

It is technically possible to move the SNF residing at all shutdown nuclear facilities in the world to remote uninhabited island locations. This course of action would, however, require access to funding, acceptance by the general public, concurrence by national governments, and approval by the International Atomic Energy Agency and, if U.S. nuclear wastes were involved, the U.S. Nuclear Regulatory Commission.

Even if all concerned should concur on this approach, project completion could easily be in the 20 to 40 year range so the first task will be to investigate one or more locations to hold only low level radioactive wastes and then offer this service to all national governments who find it of interest. Follow-up from a successful venture would expand the services offered to include SNF and high level radioactive wastes.

Thus, one possible solution is to use distance to provide for the fail safe storage of these wastes. One way to achieve this distance on an interim basis is to develop one or more remote and uninhabited island locations in the Pacific and/or Indian Oceans, where NIMBY is not a consideration.

When President Obama cancelled Yucca Mountain he set up a Blue Ribbon Commission on America's Nuclear Future to recommend an alternate approach. In an attempt to introduce this method, I submitted the below letters to the Commission:

- *Distance as an Option for the Fail Safe Interim Storage of Spent Nuclear Fuel* - 04/07/2010
- *Permanent Disposal Options* - 06/18/2010
- *32N164W* - 06/30/2010
- *Best and Final* - 07/04/2010
- *32N164W Revisited* - 07/31/2020
- *The BRC in Wiscasset, Maine on 10 August"* - 08/11/2010
- *Submarine Storage* - 08/15/2010
- *Devil's Advocate* - 08/17/2010
- *Up in the Air* - 08/18/2010
- *An Out House for Nuclear Wastes* - 08/31/2010
- *As the World Burns* - 09/30/2010
- *Problem Solved* – 02/28/2011
- *Fail Safe, Fail Safe, Fail Safe* – 03/30/2011
- *A Gadfly in the Ointment* – 04/06/2011
- *Rest In Peace* – 04/17/2011
- *Looking Under Rocks* – 05/04/2011
- *Stranded at MIT* – 05/11/2011
- *Little Women* – 05/18/2011
- *The Elephant in the Room* - 08/14/11.

These letters, along with well over 2,000 letters submitted by others to the Commission, are available for review on the web site at **http://brc.gov/**. For quicker access by those who might want to review one or more of the above letters, they are in the Files for free

downloading on the Yahoo Group for DEI at
http://tech.groups.yahoo.com/group/DirigoEnergy/.

The majority of the people submitting comments are strongly opposed to any radioactive wastes in their back yard. Let us hope that the Commission will be able to recommend a solution to this over 50 year quest to find a location for the interim storage and then permanent disposal of U.S. spent nuclear fuel and high level radioactive wastes.

LJN Enterprises will propose an alternate path with Remote Island Waste Management, hereinafter referred to as RIWM, a company being investigated to use distance as the solution to radioactive waste interim storage and/or final disposal.

The mission of RIWM will be to develop a sufficient number of remote island locations so as to provide all national governments with the option to store and/or dispose of all of their radioactive wastes.

In that they offer very few barriers to entry, the initial concentration will be on low level radioactive wastes (Class A – includes 95% of all low level radioactive wastes and requires 100 years of isolation, Class B - 300 years of isolation required, and Class C – 500 years of isolation required). After remote island sites have been developed for this purpose, they will be expanded to include high level radioactive wastes and SNF – requiring final disposal to include hundreds of thousands of years of isolation.

The ideal island type being sought for interim storage will have a lagoon (natural or artificial), an abandoned but still operational airfield for access (A private airfield is not an absolute necessity but it will facilitate the periodic crew exchange process.), and no human inhabitants. Most locations of this type are restricted to wildlife but RIWM will pursue standard operating procedures such that its presence will be a boon rather than a burden to these life forms.

There will be a trust fund set up to ensure money will be available to implement a decommissioning plan. One design requirement will include a provision similar to camping in any U.S. National Park - leave it as you found it.

Initial uninhabited locations to be researched include, but are not limited to:

- Johnston Atoll
- Wake Island
- Midway Atoll
- Salomon Islands.

The first three are in the Pacific Ocean, U.S. owned, and with natural lagoons and existing airfields. The Salomon Islands are in the Indian Ocean, UK owned, with a natural lagoon and access from an airfield on Diego Garcia – also UK owned.

Many other island nations in the Pacific and Indian Oceans will be investigated to determine if they have uninhabited islands with natural lagoons and airfield access. Perhaps this investigation will include one or more of the Solomon Islands, a sovereign nation of some 1,000 islands and 500,000 people located in the Pacific Ocean.

If an acceptable location with these criteria cannot be found, a backup position will be a remote, uninhabited island with airfield access and amenable to the creation of an artificial lagoon. One of many possible options would be the U.S. owned Jarvis Island in the Pacific Ocean with airfield access from the Republic of Kiribati's Kiritimati Island.

Seed money is necessary to incorporate RIWM, open and outfit an office and hire a nucleus group with the expertise necessary to prepare a business plan sufficiently detailed to acquire the money necessary to continue with startup and estimate the scope of storage requirements, determine how transport will be done, and evaluate functions such as manpower, logistics, warehousing, equipment, maintenance, food storage and service, etc.

Additional funding will be needed to expand the nucleus group with the personnel necessary to enlist members of a Joint Venture team to consist of expertise in the following:

1. An architect-engineer for a breakwater to serve as a source of power to a selected location and create an artificial lagoon, if required, at that location

2. An architect-engineer for a platform floating in a natural lagoon, or in an artificial lagoon created by a breakwater, on which will be stored low level radioactive wastes

3. An architect-engineer and constructor to outfit one or more selected locations to accept for extended interim storage low level radioactive wastes

4. Company or companies to construct/manufacture breakwaters and floating platforms

5. Company or companies to provide for transportation (truck, rail, ship, or air) for low level radioactive wastes from the country of origin to the remote island location

6. Company or companies to sign contracts with hospital facilities, nuclear power plant owners, and others on a global basis for utilization of the services offered by RIWM

7. Company or individuals to provide next level financing.

Additional tasks would be to acquire concept sketches for a floating power breakwater and floating waste storage platforms, cost estimates for the same, concept sketches and cost estimates for remote island outfitting, startup expenses, transport and storage rates needed for profitability, market research to determine what potential customers are paying now, pro forma financial statements to show profitability, acquire the first contract for interim storage of low level radioactive wastes, and acquire financing to continue with RIWM startup.

More funding will be needed to expand the staff, outfit a prototype remote island location for storage of low level radioactive wastes, and acquire contracts from more national governments for storage of their low level radioactive wastes.

An Initial Public Offering will provide the funds to expand into high level radioactive wastes and spent nuclear fuel.

For a Non-Disclosure Agreement and a pledge for Remote Island Waste Management send a request via email to info@dirigoenergy.org. Alternatively, join the Yahoo group at http://tech.groups.yahoo.com/group/DirigoEnergy/, go to the Files, download the pledge, and then leave the group.

The next chapter will give more information on power breakwaters.

7. POWER BREAKWATER™ ENTERPRISES

The primary objective of Power Breakwater™ Enterprises, or PBE for short, is to create calm waters and off-grid electricity at coastlines around the world.

A breakwater, either fixed or floating, will not only create calm waters but also serve as a platform on which to hold the power process equipment needed by regenerative energy sources designed to extract power in some manner from ocean wind, waves, tides, and currents.

Potential markets would include, but not be limited to support for a floating Micro-City™ discussed in chapter 3, providing for protection from the sea for the electrical generating and desalination barges of Allied Modular Power Systems discussed in chapter 5, and to serve as the power source for island locations such as those used by Remote Island Waste Management as discussed in chapter 6.

In addition, PBE can use a floating power breakwater to surround an OTEC complex as outlined in chapter 1.

One other option deserves a closer look. A heavy lift ship can move both a floating Power Breakwater™ and a floating Micro-City™ between, for example, Key West and a location on the Kennebec River in Maine so as to optimize exposure by the residents to ideal weather conditions in the winter and summer months.

Alternatively, a Micro-City™ Bed & Breakfast could use this method to relocate between comparable locations so as to maximize occupancy rates on an annual basis.

Seed money is necessary to assemble a Joint Venture team that might consist of:

- Breakwater design
- Ocean wind energy source
- Ocean wave energy source
- Ocean tidal energy source
- Volume production
- Marketing.

This Joint Venture team will produce the materials that are expected to result in customers on a global basis.

Seed money will be used to incorporate PBE, open and outfit an office and hire a nucleus group with the expertise necessary to define many energy options for a breakwater and prepare a business plan sufficiently detailed to acquire the money necessary to continue with startup.

Additional financing will be required to expand the nucleus group with the expertise necessary to design and prepare detailed drawings and cost estimates for a prototype and a team of Joint Venture participants to support it.

More funding will then be needed to expand the staff, build Power Breakwater™ prototypes to support Joint Venture agreements with

Micro-City™ Enterprises (a Snow Bird™ Bed & Breakfast moved by heavy lift ship between locations on a seasonal basis, for one example), a power source for Remote Island Waste Management, and a gas turbine/desalination plant for Allied Modular Power Systems.

For a Non-Disclosure Agreement and a pledge for Power Breakwater™ Enterprises send a request via email to info@dirigoenergy.org. Alternatively, join the Yahoo group at http://tech.groups.yahoo.com/group/DirigoEnergy/, go to the Files, download the pledge, and then leave the group.

The next chapter introduces the final business being proposed.

8. LEON'S PEERLESS PRODUCTS

The year was 1974. The place was Dardanelle, Arkansas. Suddenly, in an early December dark and stormy night, I had a lightning insight into a new way to organize a franchise business network.

In statistics, a normal distribution, commonly depicted as a bell curve, reflects an alignment in which, for a large group of people involved in the same activity, almost all are in a middle section with decreasing numbers on the right and left of the middle. My idea was to set up a franchise company that took advantage of the competitive nature of people to be number one and sprinkle that with a few incentives (pay, praise, and preferential advancement) such that employees and franchisees on the far right (the winners) would help those on the far left (the losers).

The reference to left and right has no political connotation. The bell curve under discussion simply portrays employees and franchisees on the right of center performing better than the average and those on the left below average.

If profits, customer satisfaction, and employee satisfaction are the three measured performance parameters, setting up this win-win competition to cause those on the left to increase their performance such that they move towards the right will shift the center of the bell curve to the right to ensure continuous employee and franchise improvement and lead inexorably to long term business success.

For a few details, all those in the middle of the bell curve (the first standard deviation) will control the company through voting for executive advancement and termination decisions and will receive standard performance bonuses, those in the second standard deviation at the low side will receive no bonus and at the high side double the standard bonus, those in the third and higher standard deviation on the low side will be subject to termination and at the high side triple the standard bonus plus will assist those on the low side to increase their performance. The highest bonuses and advancement options will be given to the high performers who are the most successful in helping the low performers to improve.

This basic concept evolved into an automatic control system for the large scale organization of small business which I called, at different times:

- Participative Management by Exception - the exceptionally good participate in the process by helping the exceptionally bad

- The COIL Concept - where COIL is an acronym for Coalition of Independent Lilliputians

- The Noel Notion - Noel, the opposite of my name, implies a Christmas atmosphere in which companies share information, the opposite of standard business practices

- Internal Synergistic Competition - there is competition but it is internal to the franchise network and organized such that it becomes a win-win contest

- The Theory of Business Relativity - rewards are given as a function of performance relative to the average

- Relativistic Organizational Design - advancement and termination decisions, based on performance relative to the average, are made by the average performers.

I became convinced that my best chance to test the theory would be to start a company that operated under this organizational theory from the beginning. The only option open to me at the time was to create some original products and build a business based on them. I started down this path but other priorities forced me to abandon the effort. Now, however, I have time available to try once more.

I recognize that, compared to the other ventures above, this is of a frivolous nature but its success would ensure money would be available for DEI, so I am including it as the sixth and final company in the mix.

I prepared conceptual designs for two product lines – one under the logo of **I HAVE BEEN THINKING™** by **John Anthony** and another under the logo of **WHEN YOU'RE DARING ENOUGH TO GIVE A MERRY JEST™** by **J. J. Anthony**. These pseudonyms were derived from my middle name (John) and the middle names of three of my brothers (Anthony, Joseph, and Jerome).

I am submitting the below information so as to give a better understanding of these product lines. One word of explanation - I spent seven years in the U.S. Navy, during which time I learned to cuss like a sailor. The setting permits me to do this without, hopefully, offending the ears of the faint hearted. President Nixon, also Navy trained in the use of colorful language, solved this problem with "expletive deleted" – I do it with "the noise from the airport."

Imagine an idyllic country scene. A small house on a bluff overlooks a pasture in which horses and cows are grazing peacefully and a child is cavorting with a dog. A biplane is taking off from a rough runway built into the pasture.

(THE NOISE FROM THE AIRPORT)

Imagine an international airport now surrounds the same house in the same position on the same bluff. Lines of jumbo jets are simultaneously taking off from, and landing on, multiple runways.

NOTE: The below information was recovered from a secret tape recording system installed and maintained by J. J. Anthony.

J.J.: Congratulations on your graduation, Johnny, but school's now out and the party's over! Your source of free support is hereby terminated so get off *(the noise from the airport)* and get a job!

John: Please be aware that I am now a college graduate and, as such, I would much appreciate it if, in the future, you bestow on me the respect that is my due and address me as John.

J.J.: All right! All right! Don't get in such a huff - John it will be! And just to cement this new relationship, you can stop calling me *(the noise from the airport)* and address me as J.J.

John: Now that that is established I would like to take this opportunity, J. J., to thank you very much for being so kind and generous in the past. It was truly magnanimous of you to spend all of that money on my education in accordance with the exact terms and conditions of father's will. But, now that you have brought up the subject of support, I hereby assert my long awaited independence and unequivocally state my firm intention to *NOT* get a job. As you well know, I have satisfied all conditions to now claim my portion from father's sale of Anthony Estates. Money will never be a consideration in my life so I hereby vow to never work for anyone else and fully intend to spend my legacy on my own initiatives.

J.J.: True, your share will keep you out of the poor house but, unless you leverage your inheritance many times over, it'll not put you in beachfront mansions, stretch limousines, luxury yachts, and trophy wives.

John: A business venture I have in mind might start me in that direction.

J.J.: I'm a gambler so perhaps I'll invest part of the money needed for its development. Tell me about it.

John: The opening gambit will be to publish a fairy tale that I call *THE LAND OF LO*™.

J.J.: On second thought, I might not put up a dime. If you expect to make a killing on a fairy tale, you're living in (the noise from the airport).

John: First listen and then criticize - if you must. It goes something like this:

The Land of Lo™

There is a place, way up in the sky,
It is a beautiful place; it would boggle your eye.

(Note: A joint low flying demonstration by the Blue Angels and Thunderbirds caused this extended tape gap.)

Just plant this Land of Lo™ fable;
In that place inside where everything is stable.

In that deep, dark, secret place where no one else can go;
And let it grow.

(Note: A final fly by caused more garble.)

And you will help the Whatevers in the Land of Lo™!

(Note: The complete fairy tale is included at the end of this Chapter.)

So, give me your honest reaction – what do you think?

J.J.: **It seems a bit different but confession is good for the soul so, tell me, straight from the heart, where did you steal these ideas?**

John: **From that deep, dark, secret place within me, where no one else can go.**

J.J.: **You can take that deep, dark, secret place and cram it up** *(the noise from the airport)*. **Furthermore, it'll cost too much to illustrate and then it might not sell two copies. So, if this is it, you can forget about any (noise from the airport) from me for the development of your bright ideas.**

John: **Before making any decisions you might later regret you should consider a few of my other ideas. For example, every parent and grandparent wants to cherish memories from their offspring. Similar to the ever-popular practice of copper plating a baby's shoes, copper plating their diapers would forever keep the memory of infancy fresh. Although the idea is in the public domain, trademark and copyright protection will be available if a plant hanger is called** *BABY TEARS™* **and sold with the following card:**

Baby pictures and copper shoes,
Help to drive out old age blues.

For mothers, fathers, and grandparents too,
Are always looking for memories new.

A different way to recall the past,
To make the love and affection last.

This can both hold your plants,
And mark baby's growth into training pants.

Plated with copper to form a pot,
It is a gift to cherish a lot.

For the diapers that covered those little rears,
Comes filled with the greenery called baby tears.

_J.J.:_ John - I hope you're not serious! That's the *(noisiest)* idea I've ever heard in my life. Cleaning the diapers and chrome plating the kid's turds would probably sell as well. In fact, as a crappy counter proposal you could do that very thing, place them in a stepping stool, call it *THE TURD STOOL*, and throw in this rhyme:

Baby's first use of the pot,
Is a memory to cherish a lot.

Those first little turds are plated in chrome,
Embedded in a stool and kept at home.

It's more than just a memento neat,
But a step to reach the toilet seat.

The kid can use it to take a crap,
So nature's call will not cause a flap.

No more pampers for baby to be messing,
This stool for stools is quite a blessing.

Throw out the diapers - isn't it swell,
Free at last from that awful smell!

<u>John:</u> All right! All right! I concede! I will put *BABY TEARS*™ on the back burner. But consider these two sketches. Each presents a new view to basic ideas that have been presented by others many times in the past. They could be placed on drinking glasses, on coffee mugs, or even on greeting cards.

<u>J.J.:</u> Now this has potential. There'll always be a market niche for sincere BS presented in a humorous manner. On a volume basis we should easily be able to buy low and sell high so as to make *(the noise from the airport)*. But we're going to need more than just this.

<u>John:</u> I have an idea for a simple product. The ROUND TUIT has been available for many years. It is nothing more than a circular item with a thought for procrastinators that, at last, they will be able to get some work done because they have finally gotten around to it. But it is well known that nothing ever gets completed unless a person sticks to it. So I propose to sell a paint stirrer entitled *STICK TO-IT: A Stirring Reminder*™ with the following words imprinted on it:

IT'S EASIER SAID THAN DONE all know to be true,
So if you fail at first do not quit and be blue.

The sun may be hidden and the clouds look bleak,
But never, ever give up - just continue to seek.

Examine your efforts, be they many or a few,
Then stir things around and blend them anew.

Try once more ------- and now you can do it,
For this time you have your own STICK TO-IT!

I will present the same thought in a similar fashion by appending these words to a tape measure called *THE STICK-TO-IT MEASURE*^TM

GRIN AND BEAR IT ------- ROLL WITH THE PUNCHES,
Are two ways of saying THERE ARE NO FREE LUNCHES.

IF AT FIRST YOU DON'T SUCCEED ---- TRY, TRY AGAIN,
Is another cliché that makes failure seem like sin.

So when you're beaten and bruised and down in the dumps,
Remember the strong are expected to take their lumps.

Now get back to your task; STICK-TO-IT with mirth,
And let all admire the measure of your worth!

Finally, the design science revolution promoted by the late *R.* Buckminster Fuller teaches that technology is learning to create wealth by doing more and more with less and less. This is proceeding at such a pace that, in theory, all people on earth will soon be able to live like millionaires. To publicize this thought, a ruler - colored gold - will be called *THE STICK TO-IT RULE*^TM and sold with the following words etched on it:

DO UNTO OTHER says The Golden Rule,
In the past, alas, this was only for school.

"Not enough to go around" was the common belief,
This "us or them" thought led to war and grief.

But science has now banished the validity of such lore,
And the limits of yore are with us no more.

All people on earth can now each have their druthers,
Just stick to this truth and treat all others as brothers.

__J.J.:__ Now we're really getting into the area of practicality. All three can be built cheaply and sold dearly. They'll certainly return our investment but it'd be much better if we had a more expensive product.

__John:__ Here is one that might work. A wall plaque could have various animal shapes surrounding a verse extracted from Walt Whitman's *LEAVES OF GRASS*.

To close out my present repertoire, a mirror is sold with a copy of an original painting above it. The design of the mirror is such that the picture can be removed easily and replaced with a personal photo such as wife, wedding, child, family portrait, etc. It is called **MIRROR *ON THE WALL*™** and comes with this card:

Humpty Dumpty had this mirror on his wall,
And it was the cause of that famous great fall.

The reflection of his shape was no great surprise,
But the artistry above it led to his demise.

An original design, a true work of art,
He got up to buy another for his new sweetheart.

The store, he knew, was selling out fast,
And he began to rush so he would not be last.

(Short pause in memoriam.) ***

The merchants vowed his death would not be in vain,
And quit selling this gift to keep customers sane.

For a very long time this mirror was lost --- Alas!
Now it's been saved by John - a young lad with class!

Mirror, mirror - you will be hung on the wall,
As a gift to my own "fairest of them all!"

*** Disclaimer: This is pure conjecture; it could have happened this way but there is no historical evidence that these were the actual conditions leading to Mr. Dumpty's death.

John: So my plan is to establish an identity by developing these ideas. My first objective is to devise a method to include them all under a single banner. The way I intend to do this is by using a logo of a small child sitting in the classic pose of The Thinker and then raising his arms as he has an idea. The product line will be called *I HAVE BEEN THINKING*™ and will be symbolic of a product that expresses a new thought, or at least a new way of thinking about *(the noise from the airport).*

J.J.: I must confess that I really like several of your product concepts. Let me think about this budding business a little more and then I'll get back to you.

----------------------*LATER*--------------------

J.J.: Listen up, Johnny – I mean John! The thing you are missing is a counterpoint to the thinker position. Why not try a separate line of opposite products? We could introduce a logo of a clown - The Court Jester, if you will - rolling with laughter on the back of the Sphinx. These items would be humorous, perhaps with a riddle or even an earthy tone, and could be sold under the banner of *WHEN YOU'RE DARING ENOUGH TO GIVE A MERRY JEST*™.

For one quick example, your stick to it theme can be modified for use as a blunt hint to unruly kids when developed in the form of a ruler called *STICK TO-IT: The Scold 'em Rule*™ and sold with the following card:

SPARE THE ROD AND SPOIL THE CHILD is a saying of old,
Without this rule kids forget to do as they are told.

In times of trouble when a rod would be dandy,
This rule on the wall will come in right handy.

So when you are given a simple chore or two,
Stick to the job, your excuses will NOT DO!

And ------- my child ------- you BETTER remember to do it,
Or your BUTT will be bruised with this little STICK TO-IT!

<u>John</u>: This would have to be made from foam rubber or a class action child abuse lawsuit might quickly eat up the profits.

<u>J.J.:</u> No *(noise from the airport)* problem. Especially since foam rubber rulers can be built and shipped cheaper than wooden ones.

And here are two ideas suitable for placement on drinking glasses. These could be used in conjunction with your caricatures and, by coming up with one additional idea for each, sold as an *ANTHONY SIX PACK*™.

But I'm just getting started. My next idea was inspired by a story about a famous TV personality. In his travels away from home, when in the company of many beautiful and willing show business women, he is said to always have remained true to his wife. The major reason for his faithfulness could well have been *(the noise from the airport)* his wife kept under her pillow. This product expresses that approach through *THE THING REMOVER*™.

69

Also, I believe there is a place in the peg game market for a new product. To my knowledge an old, old game called NIM has never been exposed to the mass market. In this game, fifteen playing pieces are placed in five rows in a pyramid shape with one on the top, two in the next row, then rows of three, four, and five. Players take turns with each removing as many pieces as desired but they must all come from a single row. The winner is the one who takes the last piece. A solution worked out by Charles Boulton way back in 1902 shows that the person going first can always win. The method requires a rudimentary knowledge of binary mathematics with *(the noise from the airport)* guaranteeing a victory. My plan is to market this as a con game in which the sucker going first will always lose. On a playing board shaped like a lobster, fourteen white marbles can be placed in a 5-4-3-2 triangular array with a single black marble placed on top. The game can be called *THE LOBSTER FLOP*^{TM} and be sold with this rhyme placed on a card:

Four separate rows in a game for two,
Who goes first? Decide! Please do!
Now remove some marbles - one or a few.

A single rule that you need know,
These marbles must come from only one row.

In turn now, the other does the same,
So it goes 'til the black one does remain.

If you pick up this you'll angrily hop!
For that's no marble - it's a lobster flop!

Don't punch your opponent or tell him to cram,
Don't kick his shins or holler --- DAMN, DAMN, DAMN!

Just show some manners, show some class,
Let the winner go first, then WHIP HIS ASS!

Note: In Maine, anyone can sell lobsters but only very, very rarely is an individual ingenious enough to profit from their droppings. I humbly submit to you that I am such a person. -- J.J. Anthony

For a final offering I have rethought the concept of the training stool I earlier proposed in jest. A stool can be designed to abut against the toilet. When nature requires *(the noise from the airport)* the user raises the toilet seat, mounts the stool, and assumes a natural squatting position. The stool can be called *THE HILLBILLY HELPER[TM]* and sold with the following advertisement:

ABORIGINES DO IT

(Show several pictures of aborigines in the natural squatting position they use for the elimination of bodily wastes.)

DOCTORS DO IT

(Use quote from one of Dr. John C. Lilly's books in which he recommends that one "...lift the lid and let fly.")

HILLBILLIES DO IT

(Show a hillbilly sitting in a squatting position in an outhouse.)

TRY IT - - - YOU MIGHT LIKE IT

THE HILLBILLY HELPER[TM]

<u>John:</u> You know, as strange as it seems, this is actually a very good idea. The standard construction of a toilet forces one into an unnatural position for the elimination of solid bodily wastes. I once met a hillbilly who swore that, when nature requires one to exercise the bodily function under discussion, squatting in an outhouse was much the easier method.

(Sound of a door bell ringing)

71

J.J.: **Come on in, Leon.**

Leon: **A very impressive house you have here, J. J., but the setting is quite unusual.**

J. J.: **Unusual, yes, but necessary.** *(The noise from the airport)* **was the fourth Anthony generation to occupy his family estate but he was land rich and money poor so he sold the entire 10,000 acres to developers except for this one acre parcel, on which he built the official Anthony residence. He then put a provision in his will such that, for the next three generations, the Anthony Clan would be required to live here. This necessitates that John and I have this as our permanent residence for the rest of our lives. Of course, we have made many improvements to make it livable in this location and also have several vacation homes in quite a bit more desirable settings.**

But introductions are now in order. John, please meet this old friend of mine, Leon Neihouse. He likes to putter around with new product ideas and our discussions last week resulted in the Court Jester product line I just introduced to you.

John: **Pleased to meet you, Leon. Any old friend of J. J.'s is a new friend of mine.**

Leon: **Nice thought, John, but I only met J. J. last week. The reference to "old friend" simply means I am old - up in the septuagenarian territory.**

J. J. especially liked one new product idea of mine. I am looking for something that will achieve the popularity of Pet Rocks or the Hula-Hoop and I am hoping that an adaptation of an old peg game called NIM will achieve this result. I plan to make exact facsimiles of fifteen different rocks taken from the moon and use them on a moon shaped playing board with the obvious name for the game being *MOON ROCKS*ᵀᴹ.

John: That sounds suspiciously like a product idea J. J. just introduced to me.

J.J.: Now John, I never said that that idea was original with me. In fact, I want to use this set of circumstances as a clear learning lesson to emphasize that I have not an ounce of discrimination in me. Age makes absolutely no difference to me - I'll steal a good idea from anyone, no matter how young or old they are.

Leon: (The noise from the airport) told me of the *LOBSTER FLOP*™ adaptation and I agreed immediately with the understanding that this would serve as my ante into your budding new product design group.

For another product idea I call *Freeze Frame*™, I am aligning the planets on a particular day with this placemat example showing Independence Day. It would, however, work better if several different solar system alignments would be offered as a dining room table placemat group, with one option being the four faces on the mountain.

J.J.: Leon filled me in on a bit of his background and one aspect that I found impressive was his service in the U.S. Navy aboard (the noise from the airport), a ballistic missile nuclear powered submarine that was, at the time, the most powerful warship in recorded history. I propose that we include a third product line to consist of his product designs.

Leon: I can live with that; the first two products in this line would be *Moon Rocks*™ and *The Rushmore Collection*™.

John: It is settled, then. To reflect the unique nature of what we are offering, I suggest we defer to age and call the new business *Leon's Peerless Products*. J.J., do *you concur*?

J.J.: (The noise from the airport) rules.

Leon: I agree - on one condition. All profits will be split three ways.

John: I concur - What about you, J.J.?

J.J.: A done (*noise from the airport*). So now the three of us must go out into the world and turn *Leon's Peerless Products* into a household name. Let's meet here in one week and brain storm on how we might accomplish such a goal.

----------------------*LATER*---------------------

Leon: Let's get started - I'm anxious to discuss methods and procedures to turn us all into multi-millionaires.

John: (*The noise from the airport*) and I already are multi-millionaires. My motive for getting involved in a search for even more money is to follow the lead of Ted Turner, have fun while earning several billion dollars, and then give it all away.

J.J.: You can take that (*noise from the airport*) where the sun don't shine. My reason for joining this venture is much more understandable. I want to ascend to the "Heaven on Earth" position of making Forbes' 400 Richest List.

Leon: My financial goal is much more modest - I simply want to get to where you two already are so I'll keep in the back ground and ride quietly on your coat tails.

J.J.: Since your bank account will show far and away the greatest improvement, Leon, tell us how you recommend starting (*the noise from the airport*).

Leon: I suggest we begin with a heavy emphasis on the World Wide Web. Several of the products are so unique, especially the

Thing Remover and Loving Cups, that word of mouth should leverage a few well placed ads into *(the noise from the airport)*.

John: **That would be an excellent start. We could then sell from the web a line of made-to-order products that could include a Birthday Memento consisting of a unique solar system alignment on a placemat, pictures of household pets surrounding Walt's verse on animals, and the Thing Remover adapted to permit a woman to include a special dedication and give it as a unique present to** *(the noise from the airport)*. **If this option would have been available at the time, Lorena could have given it to John Wayne and she might then not have felt the need to bob it.**

J.J.: **The web is all right for a beginning but a great many shoppers want to touch, feel, and see a product before making a final purchase decision. We can be assured of** *(the noise from the airport)* **if we design and volume produce a standard kiosk devoted exclusively to our product lines and set them up in shopping malls.**

Leon: **So we will establish an identity through channels of distribution to include a web site and shopping mall kiosks and, from there, the world is our (noise from the airport).**

That finishes the introduction to the business. It will be named Leon's Peerless Products™ (LPP) and it will sell proprietary and other related products through channels of distribution to include a web site, a catalog, kiosks, standard retail outlets, and consignment retail outlets. LPP will have an irrevocable and perpetual requirement to donate ten percent (10%) of earnings before taxes for nonprofit uses, starting with DEI.

Seed money will be needed to incorporate LPP, open and outfit an office and hire a nucleus group with the expertise necessary to develop a few introductory products, commence sales over the web and prepare a business plan sufficiently detailed to acquire the money necessary to continue with startup.

This additional money will be used to expand the management team, develop more products, initiate a catalog operation, and build models for a kiosk, standard retail outlet, and consignment retail outlet. The latter will have an added feature of accepting the products from local independent new product developers on a consignment basis.

A private stock placement will next be sold to expand the management team, develop more proprietary products, open franchisee outlets, and conduct an advertising campaign throughout the United States. The intent of this mode of operation is to provide a continuous line of new products derived from an in-house new product department and the elevation of selected consignment products for distribution throughout the LPP network.

An initial public offering will next be sold to permit expanding outlets under a goal to have seven thousand (7,000) kiosks, seven hundred (700) standard outlets, and seventy (70) consignment outlets in the United States.

The benchmark for the standard outlets will be modeled after **Dick's Sporting Goods**, which is working on an expansion goal to have 800 outlets in operation.

Actually, the very first product I worked on was *The Land of Lo™* fairy tale introduced above. The full version follows.

> *There is a place, way up in the sky,*
> *It is a beautiful place; it would boggle your eye.*
>
> *Beds are big and bouncy, you can jump very high,*
> *Rivers have paddle boats; there are slides in the sky.*
>
> *While playing in the snow the cold will not linger;*
> *On your nose, on your toes, or on your little finger.*

Swimming pools are filled by a very special hose;
The water will not get in your eyes, ears, or nose.

But the very best thing in this most happy land,
Are the Whatevers who live there; they are simply grand.

They are different, you see;
Not like you or me.

For their suns' golden rays make all things grow;
They also give life and cause the Whatevers to glow.

No sleep is needed; they will never, ever die;
As long as they have these lights in the Sky.

So they run and skip and play in the sand;
Even sing songs with their own dance band.

That's the way it is with every Heather, Shelly, Chris, and
Sam.

Oh, excuse me! That's the way it was before that big
game in Lo Land.

Now the Land of Lo was of quite a different kind;
There were none of the good things you normally find.

For the land was covered with a light bluish haze,
Which kept out part of the suns' golden rays;
And the result of this all was to cause a slight daze.

So no one went there - except to play games.

For it really was fun and became quite the rage,
To run into the light bluish haze,
Stay for awhile until one felt a slight craze;
Then rush out again into all the suns' golden rays.

Then someone there, no one knows quite who;
Suggested a game, something different to do.

They built a great wall around all of Lo Land;
They built it with care; they built it to stand.

Then sides were chosen, a coin they did flip;
And side number two got to make the first trip.

On the inside of the wall in the light bluish haze,
Side number one built a great maze;
They built it to confuse and daze.

They put holes in the wall that were very hard to see;
When they were finished they all clapped with glee.

Side number two then the gates they went through;
Believe me, the number of Whatevers were quite a few;
Forty million, three thousand, and ninety-two.

The gates were all closed, the game it began;
They all ran about, at first without plan.

Ten million or so found their way through the maze;
But since all the suns' rays could not get through the
haze,
The rest became confused and dazed.

Then someone had a plan to help them out of their plight;
They built great machines to give artificial light.

This was a super idea, it really helped their sight;
And ten million more got out, but then, a long night.

For the effect of the machines was dreadful you know,
Their pollution caused black clouds in the sky of Lo.

Soon none of the suns' rays could get through the haze;
And without the suns' rays the Whatevers were crazed.

They even started to sleep in the strange Land of Lo;
And sleep was something which no one did know.

They then dreamed of their past glory and fame;
Woke again and remembered it was only a game.

But artificial light and clouds of gray,
Caused these happy memories to go completely away.

This went on for several sleeps or so,
Until they forgot where they were trying to go.

Their friends on the outside saw it was no longer any fun;
They threw open the gates and said "Ha Ha, We Won."

But it was too late, a strange thing had taken;
The Whatevers inside had become slightly misshapen.

For they had completely lost their vibrant glow,
Now they liked living under the black clouds of Lo.

A very drab sight, no longer so beautiful to view;
Of those still trying to get out, only one or two.

Those who looked at their old friends so bright,
Would turn away dazed and soon forget that sight.

But the strangest thing of all;
Most would not look, not even at all.

They no longer ran or skipped or played games in the
sand;
They just sat around and talked in that very sad land.

A few stayed in and tried to show them the way;
Some would listen but most turned away.

Drastic steps were required, they needed a great plan,
To get their old friends from that very dull land.

So each year many return to explain about Lo Land;
And every year a few more understand.

Now the joy and happiness of those that do glow,
Absorb some of the pollution in the dark sky of Lo.

So a few of the suns' rays can get through the haze,
To cause laughter and carefreeness during all these days.

But at the end of this time the yearning is worse it seems;
All the light leaves their eyes, not a single one gleams.

For no one does remember that the artificial light
machines,
Are what produce the pollution that keeps out the suns'
beams.

And that is the problem in the strange Land of Lo.

How can the Whatevers that glow help their old friends to
remember?

At the rate it is going it will take twenty years from this
December.

Kristina, would you like to help those dull Whatevers to
glow?
You can, you know.

Just plant this Land of Lo fable;
In that place inside where everything is stable.

In that deep, dark, secret place where no one else can go;
And let it grow.

And then every day of your life, for a moment or so;
Stop and think of the Whatevers in the Land of Lo.

And to those that pass as they come and go,
Smile and give a big Hello.

This joy and happiness you create below,
Will fly into the sky of Lo;
And absorb some of their pollution for a moment or so.

Then some of their suns' rays,
Will get through the haze.

And you will help the Whatevers in the Land of Lo!

For a Non-Disclosure Agreement and a pledge for Leon's Peerless Products send a request via email to info@dirigoenergy.org. Alternatively, join the Yahoo group at http://tech.groups.yahoo.com/group/DirigoEnergy/, go to the Files, download the pledge, and then leave the group.

If only one of the six companies proposed in this book succeeds, DEI will have the financial wherewithal to be a Player on the World Stage.

That's it, folks.

Note - A summary of the information presented in this book is available on the web at http://dirigoenergy.org/.

ABOUT THE AUTHOR

Leon Neihouse was born on 10 May 1939 to the Prairie View, Arkansas farm family of Leo Columbus and Mary Elizabeth Neihouse. Their children were, in order of birth with the oldest first: Harold, Robert, Bernadine, Mary Louise, George, Norbert, Imelda, Dorothy, Constance, and Leon.

Mr. Neihouse's initial education started in the parochial school system of Fort Smith, Arkansas culminating in graduation from St. Anne's Academy. He continued at the University of Dallas and ended at the University of Wisconsin at Madison under a three year National Science Foundation Fellowship in Physics but he quit after two semesters.

He then joined the U.S. Navy, entered Officer Candidate School in Newport, Rhode Island, was accepted into the nuclear power program, attended Nuclear Power School in Bainbridge, Maryland; Nuclear Power Prototype in West Milton, New York; Submarine School in Groton, Connecticut; new construction on the USS Simon Bolivar (SSBN 641) in Newport News, Virginia; made three deterrent patrols operating out of Charleston, South Carolina during which time he qualified in Submarines and as Chief Engineer; and served two years in a staff position at nuclear power school in Mare Island, California to complete his active duty service in 1969. He was honorably discharged several years later from the U. S. Naval Reserves as a Lieutenant Commander.

During his Navy tour of duty he married Billie Sandra, a former United Airlines flight attendant. They have a son, John Parrish (now a tax attorney in Northwest Arkansas) and a daughter, Kristina Anne (now a Librarian in Key West).

After discharge from the Navy he worked three years for the Tennessee Valley Authority at Brown's Ferry in Athens, Alabama and four years for

Combustion Engineering on the start up of Maine Yankee Atomic Power Plant in Wiscasset and Arkansas Nuclear One, Unit Two near Russellville.

In his mid-thirties he became interested in business and hypothesized a way to increase long term profitability.

He started taking undergraduate courses in business during the evening so as to develop the theory in a more explicit format but became dissatisfied with progress so he quit his day job with Combustion Engineering and returned to graduate school on a full time basis to pursue a Master's degree in Business Administration.

He enrolled in the MBA program at the University of Southern Maine (USM) as a full time student but ran out of money before he could finish so he accepted a job in the Bath, Maine office of John J. McMullen & Associates, a Naval Architect and Marine Engineering firm, and continued his pursuit of a MBA on a part-time basis.

MBA support courses he had taken at the University of California at Berkeley, Arkansas Tech University, and the University of Arkansas at Fayetteville were transferred to USM and, for his final MBA course, he wrote a report entitled Relativistic Organizational Design (ROD), which provided the theoretical framework for this new approach to increasing organizational effectiveness.

He then tried many avenues, which continues to this day, but he has yet to find a method to test the theory.

He, therefore, continued to work for John J. McMullen & Associates, later purchased by Alion Science & Technology, and was employed for over 34 years serving as a Project Manager supporting the U.S. Navy in their FFG (frigate), CG (cruiser), and DDG (destroyer) shipbuilding programs.

His objective is to have all six of the businesses proposed by LJN Enterprises in this book developed using ROD techniques but he hopes that at least one will be successful to such an extent that it will serve as the vehicle for a practical test of the organizational theory.

www.ingramcontent.com/pod-product-compliance
Lightning Source LLC
Chambersburg PA
CBHW071244170526
45165CB00003B/1229